MATH DIALOGUE: FUNCTIONS

METIN BEKTAS

Copyright © 2015 Metin Bektas

All rights reserved.

ISBN: 9781520871769

DEDICATION

This book is dedicated to my family.

CONTENTS

1. Lesson I: Introduction   1

- Domain

- Application

- Plots

2. Lesson II: Linear Functions   27

- Y-Intercept / Slope

- Slope Triangle

- Roots

3. Lesson III: Quadratic Functions   53

- Constants

- Vertex Point

- Roots

- Factored Form

Lesson I

- **Introduction**

- **Domain**

- **Application**

- **Plots**

T:

What are functions? I could just insert the standard definition here, but I fear that this might not be the best approach for those who have just started their journey into the fascinating world of mathematics. For one, any common textbook will include the definition, so if that's all you're looking for, you don't need to continue reading here. Secondly, it is much more rewarding to build towards the definition step by step. This approach minimizes the risk of developing deficits and falling prey to misunderstandings.

S:

So where do we start?

T:

We have two options here. We could take the intuitive, concept-focused approach or the more abstract, mathematically rigorous path. My recommendation is to go down both roads, starting with the more intuitive approach and taking care of the strict details later on. This will allow you to get familiar with the concept of the function and apply it to solve real-world problems without first delving into sets, Cartesian products as well as relations and their properties.

S:

Sounds reasonable.

T:

Then let's get started. For now we will think of a function as a mathematical expression that allows us to insert the value of one quantity x and spits out the value of another quantity y. So it's basically an input-output system.

S:

Can you give me an example of this?

T:

Certainly. Here is a function: $y = 2 \cdot x + 4$. As you can see, there are two variables in there, the so-called independent variable x and the dependent variable y. The variable x is called independent because we are free to choose any value for it. Once a value is chosen, we do what the mathematical expression tells us to do, in this case multiply two by the value we have chosen for x and add four to that. The result of this is the corresponding value of the dependent variable y.

S:

So I can choose any value for x?

T:

That's right, try out any value.

S:

Okay, I'll set x = 1 then. When I insert this into the expression I get: y = 2·1 + 4 = 6. What does this tell me?

T:

This calculation tells you that the function y = 2·x + 4 links the input x = 1 with the output y = 6. Go on, try out another input.

S:

Okay, can I use x = 0?

T:

Certainly. Any real number works here.

S:

For x = 0 I get y = 2·0 + 4 = 4. So the function y = 2·x + 4 links the input x = 0 with the output y = 4.

T:

That's right. Now it should be clear why x is called the independent variable and y the dependent variable. While you may choose any real number for x, sometimes there are common sense restrictions though, we'll get to that later, the value of y is determined by the form of the function. A few more words on terminology and notation. Sometimes the output is also called the value of the function. We've just found that the function y = 2·x + 4 links the input x = 1 with the output y = 6. We could restate that as follows: at x = 1 the function takes on the value y = 6. The other input-output pair we found was x = 0 and y = 4. In other words: at

$x = 0$ the value of the function is $y = 4$. Keep that in mind.

As for notation, it is very common to use $f(x)$ instead of y. This emphasizes that the expression we are dealing with should be interpreted as a function of the independent variable x. It also allows us to note the input-output pairs in a more compact fashion by including specific values of x in the bracket. Here's what I mean.

For the function we can write: $f(x) = 2 \cdot x + 4$. Inserting $x = 1$ we get: $f(1) = 2 \cdot 1 + 4 = 6$ or, omitting the calculation, $f(1) = 6$. The latter is just a very compact way of saying that for $x = 1$ we get the output $y = 6$. In a similar manner we can write $f(0) = 4$ to state that at $x = 0$ the function takes on the value $y = 4$. Please insert another value for x using this notation.

S:

Will do. I'll choose $x = -1$. Using this value I get: $f(-1) = 2 \cdot (-1) + 4 = 2$ or in short $f(-1) = 2$. So at $x = -1$ the value of the function is $y = 2$. Is all of this correct?

T:

Yes, that's correct.

S:

You just mentioned that sometimes there are common sense restrictions for the independent variable x. Can I see an example of this?

T:

Okay, let's get to this right now. Consider the function $f(x) = 1/x$. Please insert the value $x = 1$.

S:

For x = 1 I get f(1) = 1/1 = 1. So is it a problem that the output is the same as the input?

T:

Not at all, at x = 1 the function f(x) = 1/x takes on the value y = 1 and this is just fine. The input x = 2 also works well: f(2) = 1/2, so x = 2 is linked with the output y = 1/2. But we will run into problems when trying to insert x = 0.

S:

I see, division by zero. For x = 0 we have f(0) = 1/0 and this expression makes no sense.

T:

That's right, division by zero is strictly verboten. So whenever an input x would lead to division by zero, we have to rule it out. Let's state this a bit more elegantly. Every function has a domain. This is just the set of all inputs for which the function produces a real-valued output. For example, the domain of the function f(x) = 2·x + 4 is the set of all real numbers since we can insert any real number x without running into problems. The domain of the function f(x) = 1/x is the set of all real numbers with the number zero excluded since we can use all real numbers as inputs except for zero.

Can you see why the domain of the function f(x) = 1/(3·x - 12) is the set of all real numbers excluding the number four? If it is not obvious, try to insert x = 4.

S:

Okay, for x = 4 I get f(4) = 1/(3·4 - 12) = 1/0. Oh yes, division by zero again.

T:

Correct. That's why we say that the domain of the function f(x) = 1/(3·x - 12) is the set of all real numbers excluding the number four. Any input x works except for x = 4. So whenever there's an x somewhere in the denominator, watch out for this. Sometimes we have to exclude inputs for other reasons, too. Consider the function f(x) = sqrt(x). The abbreviation "sqrt" refers to the square root of x. Please compute the value of the function for the inputs x = 0, x = 1 and x = 2.

S:

Will do.

f(0) = sqrt(0) = 0

At x = 0 the value of the function is 0.

f(1) = sqrt(1) = 1

At x = 1 the value of the function is 1.

f(2) = sqrt(2) = 1.4142 ...

At x = 2 the value of the function is 1.4142 ... All of this looks fine to me. Or is there a problem here?

T:

No problem at all. But now try x = -1.

S:

Okay, f(-1) = sqrt(-1) = ... Oh, seems like my calculator spits out an error message here. What's going on?

T:

Seems like your calculator knows math well. There is no square root of a negative number. Think about it. We say that the square root of the number 4 is 2 because when you multiply 2 by itself you get 4. Note that 4 has another square root and for the same reason. When you multiply -2 by itself, you also get 4, so -2 is also a square root of 4.

Let's choose another positive number, say 9. The square root of 9 is 3 because when you multiply 3 by itself you get 9. Another square root of 9 is -3 since multiplying -3 by itself leads to 9. So far so good, but what is the square root of -9? Which number can you multiply by itself to produce -9?

S:

Hmmm ... 3 doesn't work since 3 multiplied by itself is 9, -3 also doesn't work since -3 multiplied by itself is 9. Looks like I can't think of any number I could multiply by itself to get the result -9.

T:

That's right, no such real number exists. In other words: there is no real-valued square root of -9. Actually, no negative number has a real-valued square root. That's why your calculator complained when you gave him the task of finding the square root of -1. For our function f(x) = sqrt(x) all of this means that inserting an x smaller than zero would lead to a nonsensical result. We say that the domain of the function f(x) = sqrt(x) is the set of all <u>positive</u> real numbers including zero.

In summary, when trying to determine the domain of a function, that is, the set of all inputs that lead to a real-valued output, make sure to exclude any values of x that would a) lead to division by zero or b) produce a negative number under a square root sign. Unless faced with a particularly exotic function, the domain of the function is then simply the set of all real numbers excluding values of x that lead to division by zero and those values of x that produce negative numbers under a square root sign.

I promise we will get back to this, but I want to return to the concept of the function before doing some exercises. Let's go back to the introductory example: f(x) = 2·x + 4. Please make an input-output table for the following inputs: x = -3, -2, -1, 0, 1, 2 and 3.

S:

Okay.

f(-3) = 2·(-3) + 4 = -2

f(-2) = 2·(-2) + 4 = 0

f(-1) = 2·(-1) + 4 = 2

f(0) = 2·0 + 4 = 4

f(1) = 2·1 + 4 = 6

f(2) = 2·2 + 4 = 8

f(3) = 2·3 + 4 = 10

So at x = -3 the value of the function is y = -2, at x = -2 the value of the function is y = 0, and so on.

T:

Correct.

S:

May I ask: what's the point of all this? What possible uses could this function have?

T:

Excellent question. I can think of two great applications on the spot. One comes from the field of physics. Suppose x and y were physical quantities. The function would then establish a mathematical link between the two. For example, the speed of sound waves is intimately connected to the temperature of air. Wouldn't it be great to have a function where I could insert a value for the temperature and get the speed of sound as the output? I could then easily check how fast sound waves travel at any given temperature.

S:

Can you show me this function, please.

T:

With pleasure. The corresponding function is $f(x) = 20 \cdot \sqrt{273.15 + x}$ with x being the air temperature in degrees Celsius and the output $f(x)$ being the speed of sound in meters per second. Use this function to compute the speed of sound on a cold winter day, use x = -15 °C, and compare this to the speed of sound on a hot summer day, x = 35 °C.

S:

I'll try. For x = -15 I get:

$f(-15) = 20 \cdot \sqrt{273.15 - 15}$

f(-15) = 20·sqrt(258.15)

f(-15) ≈ 321

So when the air temperature is -15 °C, sound waves travel at a speed of roughly 321 meters per second.

T:

Excellent. Go on.

S:

Inserting x = 35 leads to:

f(35) = 20·sqrt(273.15 + 35)

f(35) = 20·sqrt(308.15)

f(35) ≈ 351

In air having the temperature 35 °C sound waves thus travel at around 351 meters per second.

T:

Well done! 351 is approximately 10 % greater than 321, so the exciting conclusion is that on a hot summer day sound travels 10 % faster than on a cold winter day. Note that the function f(x) = 20·sqrt(273.15 + x) does not allow us to insert values for x that are smaller than -273.15. For example, when you try to find the value of the function at x = -300, you get:

f(-300) = 20·sqrt(273.15 - 300)

f(-300) = 20·sqrt(-26.85)

A negative number under the square root sign, as you know, we cannot evaluate this abomination. But here the limitation makes sense. Absolute zero lies at -273.15 °C, no object can ever have a temperature below that. So the domain of this function, which consists of all real numbers greater than or equal to -273.15, just reflects this physical reality.

Another quick example of the usefulness of functions before we continue. Suppose you throw a stone at a 45° angle relative to the ground with the initial velocity x given in meters per second. Applying the basic laws of mechanics, we find that the range of the stone is $f(x) = 0.1 \cdot x^2$, expressed in the unit meters. So when throwing said stone at the initial speed x = 5 meters per second, the range will be $f(5) = 0.1 \cdot 5^2 = 2.5$ meters. With the initial speed x = 10 meters, the stone will travel the horizontal distance $f(10) = 0.1 \cdot 10^2 = 10$ meters before returning to the ground.

But functions are not just useful for doing physics. We can also interpret them geometrically. This is what we shall do now. Do you remember the Cartesian coordinate system?

S:

I think so. There is one horizontal line, the x-axis, and one vertical line, the y-axis. The two lines meet at the origin.

T:

That's right. Do you also remember how to find a given point in the coordinate system? Can you show me the point P(2, 4)?

S:

I can try. The first number in P(2, 4) should be the x-coordinate, the second number the y-coordinate. So to find the point, you start at the origin, go two units to the right in horizontal direction, then four units upwards.

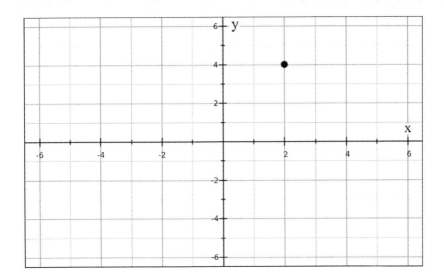

T:

Very nice. What about the point P(-5, -3)?

S:

To find P(-5, -3) we must go five units to the left from the origin and then three units downwards.

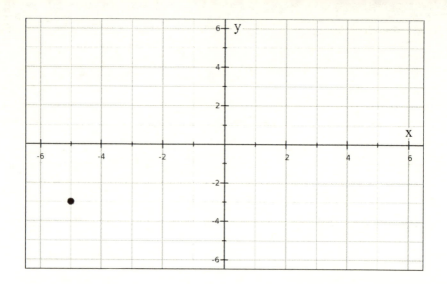

T:

Great. Now let's go back to the function f(x) = 2·x + 4. Here's the shortened version of the table we found earlier:

f(-3) = -2

f(-2) = 0

f(-1) = 2

f(0) = 4

f(1) = 6

f(2) = 8

f(3) = 10

So when we insert x = -3 into the function, we get the output y = -2, for x = -2 we get the output y = 0, and so on. This is not news. However, now let's interpret these input-output pairs geometrically. We may think of every input-output pair x and y as a point P(x, y) in the Cartesian coordinate system. So ...

... f(-3) = -2 corresponds to P(-3, -2)

... f(-2) = 0 corresponds to P(-2, 0)

... f(-1) = 2 corresponds to P(-1, 2)

... f(0) = 4 corresponds to P(0, 4)

... f(1) = 6 corresponds to P(1, 6)

... f(2) = 8 corresponds to P(2, 8)

... f(3) = 10 corresponds to P(3, 10)

In the image below you can see what it looks like when we plot all of these points in the same coordinate system.

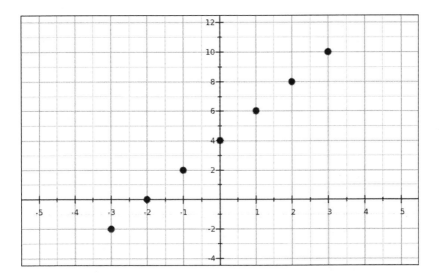

Seems like the points generated by the function lie on a straight line. To confirm this, we have to compute a lot more points though. You can do this the old-fashioned way by choosing more values of x, finding the corresponding values of y using the function and then plotting the resulting points P(x, y) in the coordinate system. But let's do it the smart way, that is, humbly asking a computer for help. For this we can go to the free website graphsketch.com and type in the function $f(x) = 2 \cdot x + 4$. Use the symbol * for multiplication. Once you are finished typing in the function, make sure to set the x-range as well as the y-range. For the image below I chose the ranges from x = -5.5 to x = 5.5 and from y = -4.5 to y = 12.5.

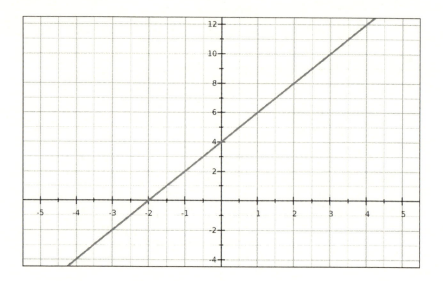

As you can see, the computer was able to produce a lot more points than we could ever have, so many that instead of a series of points you get a continuous line. This confirms our initial assumption: the graph of the function f(x) = 2·x + 4 is a straight line crossing the x-axis at x = -2 and the y-axis at y = 4.

Now it's your turn. Let's go back to the square root function, f(x) = sqrt(x). I would like you to do the following: make a table of function values for the input values x = 0, 1, 2, 3, 4 and 5, then convert all the input-output pairs into points and plot them in a coordinate system. After that, go to graphsketch.com and have it produce the graph of f(x) = sqrt(x). Compare your plot with the plot from the website.

Here's a tip: the square root of x is equal to x to the power of one-half. In short: sqrt(x) = $x^{1/2}$. So if your calculator doesn't have the square root sign, just calculate x to the power of one-half. When using the Windows calculator, press "y" for "to the power of".

S:

I'll start with the table:

f(0) = sqrt(0) = 0

f(1) = sqrt(1) = 1

f(2) = sqrt(2) ≈ 1.41

f(3) = sqrt(3) ≈ 1.73

f(4) = sqrt(4) = 2

f(5) = sqrt(5) ≈ 2.24

So the points are:

P(0, 0)

P(1, 1)

P(2, 1.41)

P(3, 1.73)

P(4, 2)

P(5, 2.24)

Here's my plot:

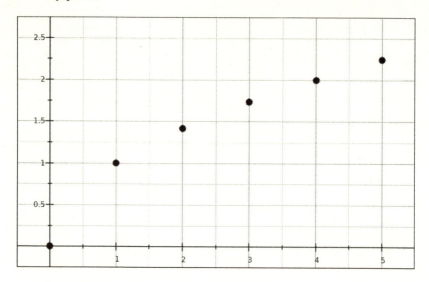

T:

Alright, it looks like all the points are in the right place. Now let's compare this to the plot produced by graphsketch.com. Use the ranges from x = -0.5 to x = 5.5 and y = -0.25 to y = 2.75. To make the result look really nice, you can set the x tick distance to 0.5, the y tick distance to 0.25 and instruct the applet to label every second tick.

S:

Okay, here's the plot from the website:

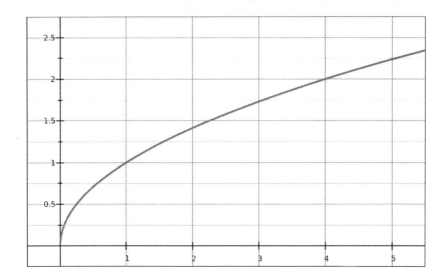

T:

That agrees with your plot very well. Fantastic! You can see that the computer is much more effective at generating the graph of a function, but you should still know how it is produced. The computer does the same thing you just did a moment ago: choose an x, insert that into the function, compute the corresponding y, plot the point P(x, y). The difference is that the computer can do this a million times per second while we are unfortunately much slower than that.

S:

Is it possible to know the shape of a function by simply looking at the mathematical expression?

T:

Good question. With experience you will be able to "see" the graph in your mind for the most common functions.

S:

What are the most common functions?

T:

The most common functions are linear functions, quadratic functions, power (or allometric) functions, exponential functions and trigonometric functions. Quite often you'll also come across cubic functions, square root functions and logarithmic functions. Of course there are many other types of functions, but for now your goal should be to get familiar with the most common ones.

This concludes the first lesson. Let's briefly summarize the core concepts. Can you tell me how we defined the function?

S:

A function is basically an input-output system. You throw a value x into a mathematical expression, do the calculation and get a value y in return.

T:

Correct. Since we are free to choose the value of x, we call x the independent variable and y the dependent variable.

S:

Can we have more than two variables?

T:

This would be a topic for a book in advanced mathematics, but I'll give a brief answer here. Yes, it is possible to have multiple inputs and / or multiple outputs. Consider the expression:

$f(x_1, x_2) = 2 - x_1 + 4 \cdot x_2$

This is a function with two independent variables x_1 and x_2 and one dependent variable we shall also call y. Every pair of values x_1 and x_2 produces one output y. A few examples:

$f(0, 0) = 2 - 0 + 4 \cdot 0 = 2$

$f(0, 1) = 2 - 0 + 4 \cdot 1 = 6$

$f(-2, 2) = 2 - (-2) + 4 \cdot 2 = 12$

How to interpret this? We can think of this function as producing a number for every point $P(x_1, x_2)$ in the plane. So the origin $P(0, 0)$ is assigned the number 2, the point $P(0, 1)$ the number 6 and the point $P(-2, 2)$ the number 12. This could be a temperature field for example or the height of the terrain at this given point. Of course we can easily define a function with even more independent variables:

$f(x_1, x_2, x_3) = 4 \cdot x_1 + x_2 - x_3$

This function has three independent variables x_1, x_2 and x_3 and one dependent variable y. You input three freely chosen values to get one output. Here's how it works:

$f(0, 0, 0) = 4 \cdot 0 + 0 - 0 = 0$

$f(1, 1, 0) = 4 \cdot 1 + 1 - 0 = 5$

$f(1, 2, 6) = 4 \cdot 1 + 2 - 6 = 0$

This is more difficult to interpret. Think of a Cartesian coordinate system with an additional axis, an axis we shall call the z-axis and which lies at a right angle to the plane defined by the x- and y-axis. We then have a three-dimensional coordinate system. To specify a point within it, you need three values: an x-coordinate, a y-coordinate and a z-coordinate. Hence, we can interpret any triple of numbers as a point in this three-dimensional coordinate system.

Accordingly, we can interpret the function $f(x_1, x_2, x_3)$ as a system that assigns each point in three-dimensional space a number. The origin $P(0, 0, 0)$ is assigned the number 0, the point $P(1, 1, 0)$ associated with the number 5 and the point $f(1, 2, 6)$ with the number 0. Again, we could think of this as a temperature field or the distortion of space into the fourth dimension at this point.

Mathematically you can easily go beyond three independent variables, but the functions get more and more difficult to interpret, at least in the physical and geometrical sense. You can also have multiple dependent variables, but since this requires knowledge in vector algebra, we will not delve into that here. I hope this answers your question. Back to our brief summary of the first lesson. What can you tell me about the domain of a function?

S:

The domain of a function is the set of all inputs we can use with the given function.

T:

Careful, this is not specific enough. Let me correct that: The domain of a function is the set of all inputs that produce a real-valued output.

S:

What exactly is a real-valued output?

T:

This means that the output is a real number. The set of real numbers consists of rational numbers, which are integers or fractions, and irrational numbers such as pi or the non-integer square roots of positive numbers. All of these are allowed as outputs of a function. Not part of the real numbers are expressions that have a zero denominator, these are just undefined expressions, or the square roots of negative numbers. Inputs that produce such "weird" expressions and numbers can not be included in the domain. For almost all functions you can think of the domain is just the set of all real numbers excluding those inputs x that lead to a zero denominator or to a negative number under the square root sign.

S:

Okay, got it. To determine the domain I start with the set of real numbers and exclude all inputs that lead to a zero denominator or negative numbers under the square root sign.

T:

Exactly. Onto applications. Can you tell me what real-world uses there are for functions?

S:

Functions can connect two physical quantities, allowing you to calculate the one from the other. The example we looked at was the function $f(x) = 20 \cdot \text{sqrt}(273.15 + x)$ which allows us to insert the air temperature x and calculate the speed of sound y at that temperature.

T:

Yes, but make sure to always include the units. In the above case the temperature must be inserted in degrees Celsius and the output will be in meters per second. Okay, now tell me how I can plot a given function by hand. Suppose you are given the function $f(x) = x^2$ and want to make a plot.

S:

I first have to choose some values for x.

T:

Give me an example.

S:

Okay, I'd choose x = -2, -1, 0, 1 and 2.

T:

That's not going to be a thorough plot, but for the brief summary it's fine.

S:

Then I produce a table of function values:

$f(-2) = (-2)^2 = 4$

$f(-1) = (-1)^2 = 1$

$f(0) = 0^2 = 0$

$f(1) = 1^2 = 1$

$f(2) = 2^2 = 4$

And plot the corresponding points:

P(-2, 4)

P(-1, 1)

P(0, 0)

P(1, 1)

P(2, 4)

T:

Okay, I can see you understood everything. Great job. We are done with lesson one now. Lesson two will be a detailed look at the most basic and most important class of functions, the linear functions. Before beginning the lesson, you should know how to solve linear equations. For this, you can either download my free book "Algebra - The Very Basics" from Amazon or you can watch carefully as I show you an example here. Since this book is about functions and not equations, I will leave it at this one example though.

Let's say we want to solve:

$7 \cdot x + 19 = 40$

The equation says: seven times some number x plus 19 is equal to 40. Which number satisfies this condition? The goal is to bring the equation to the form:

$x = ...$

To do this, first subtract 19 from both sides:

$7 \cdot x + 19 = 40 \quad / -19$

$7 \cdot x = 21$

Then divide by 7:

$7 \cdot x = 21 \quad / :7$

$x = 21 / 7 = 3$

So the solution of the linear equation is x = 3. Okay, that was it. See you soon for lesson number two and make sure not to forget the core principles we have discussed in lesson one.

Lesson II

- **Linear Functions**

- **Y-Intercept / Slope**

- **Slope Triangle**

- **Roots**

T:

I see you are back for more. Then let's get started. I printed out the plot of three functions for you. Have a look at them and try to find some sort of pattern.

$f(x) = 3 \cdot x - 2$

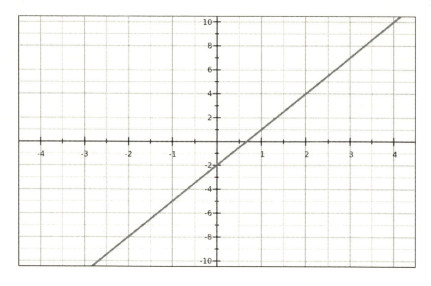

$f(x) = 2 \cdot x + 1$

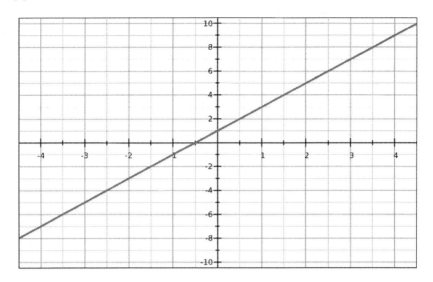

$f(x) = -2 \cdot x + 5$

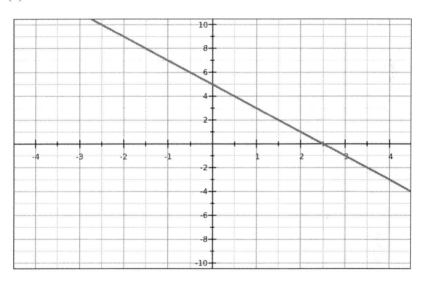

Did you notice something? Let's start with the obvious. All of these functions have the general form:

$f(x) = a \cdot x + b$

With two fixed constants a and b. Functions of this form are called linear functions. What can you say about the shape of the graph of linear functions?

S:

It seems that the graph is always a straight line.

T:

Yes, this is indeed true. The graph of all linear functions is a straight line. Did you notice something else? Have a close look at the constant b in the example.

S:

It seems that the graph always crosses the y-axis at y = b. For example, the graph of the function $f(x) = 3 \cdot x - 2$ crosses the y-axis at y = -2 and the graph of $f(x) = 2 \cdot x + 1$ crosses the y-axis at y = 1. The third example confirms this pattern.

T:

Great observation. And this makes sense. To find the value at which a function crosses the y-axis, we must insert x = 0, that is, we must find f(0). For linear functions we get:

$f(0) = a \cdot 0 + b = b$

So at x = 0 the value of the linear function is y = b. In other words: the graph goes through the point P(0, b). This point clearly lies on the y-axis. So we'll keep in mind: the constant b immediately tells us where the straight line crosses the y-axis. For this reason the constant is also called the y-intercept or initial value. Can you find a pattern for the other constant as well? This is a bit more difficult.

S:

Hmmm ... I see that when the constant a is positive, as it is the case in the first two plots, the function increases as we go to the right. But when the constant a is negative, the function decreases when going to the right.

T:

Yes, you are absolutely right. But careful with the terminology here. A function never increases or decreases. The function is a mathematical expression and it thus does not make sense to say that it increases or decreases. What you should say is that when a is positive, <u>the value of the function</u> increases as we go to the right. And when a is negative, <u>the value of the function</u> decreases when going to the right.

Enough with the terminology, back to the graphs. Have another look at the first two example graphs. Note that the first graph is steeper than the second. While the value of the function $f(x) = 3 \cdot x - 2$ goes up by three units when going to the right by one unit, the value of $f(x) = 2 \cdot x + 1$ goes up by only two units as we go to the right by one unit. Do you see that?

S:

So the constant a tells us by how much the value of the function goes up as we go one unit to the right?

T:

Up or down, depending on the sign. Look at the third example graph. Here the constant a has the value a = -2 and the graph goes down by 2 units as we go to the right by one unit. So yes, you got it right. The constant a in the linear function f(x) = a·x + b tells us by how much the straight line goes up or down as we go one unit to the right. For this reason a is called the slope of the function. The larger a is, the steeper the line. Confirm that we properly understood the meaning of the constants a and b by looking at the graph below.

f(x) = 4·x - 6

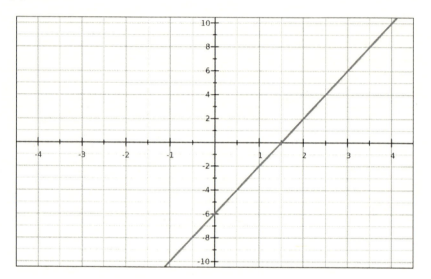

S:

Okay, the constant b has the value b = -6. This is where the line should cross the y-axis. The graph confirms this. The slope has the value a = 4, this means that the graph should go up by 4 units as we go to the right by one unit. Yes, this is the case. At x = 0 the function has the value y = -6 and at x = 1, one unit to the right, it has the value y = -2. So it went up by 4 units.

T:

Very nice. This knowledge allows you to make a quick plot of any linear function. What do you think the graph of the function $f(x) = -3 \cdot x + 2$ will look like?

S:

Well, it should begin at y = 2.

T:

Be a bit more careful with the terminology. Technically the graph does not begin or end anywhere. It spans from minus infinity to plus infinity. Please include the corresponding x-value when talking about a certain value of y.

S:

Okay, what I meant was that at x = 0 the function has the value y = 2. So it crosses the y-axis at y = 2. When you go from this point one unit to the right, the graph should go down by three units.

T:

Correct, that's all you need to draw the graph. And here it is in all its glory, the graph of $f(x) = -3 \cdot x + 2$.

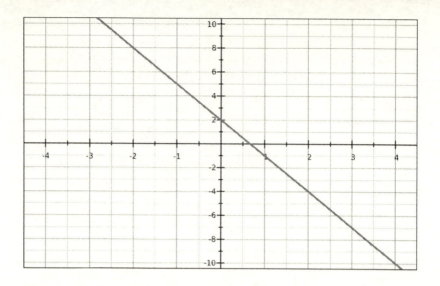

Now let's reverse the situation. I'll show you two graphs and ask you to determine the corresponding linear functions. Knowing what you know now, this shouldn't be too difficult. Start by noting the y-intercept b, then find the slope a and finally write down the function. Here's graph number one:

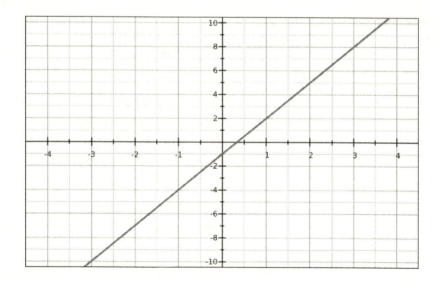

S:

The function seems to cross the y-axis at y = -1. So the y-intercept is b = -1.

T:

That's right, go on.

S:

The value of the function changes from y = -1 to y = 2 when going from x = 0 to x = 1. That's three units up, so a = 3. The corresponding function should thus be f(x) = 3·x - 1.

T:

Very nice! Now onto graph number two:

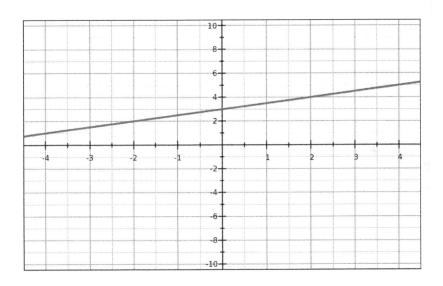

S:

Okay, the y-intercept is b = 3. The function seems to go up by only half a unit when going to the right by one unit. Does that mean that a = 0.5 is the correct value for the slope?

T:

Correct.

S:

So the function is f(x) = 0.5·x + 3.

T:

Right again. As you can see, when the slope is smaller than one, the strategy of reading the slope off by going one unit to the right might lead to problems. In this case it is much better to go to the right by more than just one unit and apply the general formula for calculating the slope: the slope a is equal to the change in y divided by the change in x. In short:

a = Δy / Δx

The Greek letter Δ (delta) stands for "change in". You should definitely keep that formula in mind. Let's go back to the previous graph. You will notice that I drew in a triangle that will make it easier to read off the changes.

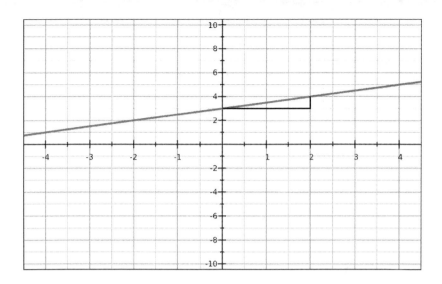

From the slope triangle we can see that the value of the function goes up by one unit as we go two units to the right. Hence, a change in x by $\Delta x = 2$ units corresponds to a change in y by $\Delta y = 1$ unit. This leads to the slope:

$a = \Delta y / \Delta x = 1 / 2 = 0.5$

In agreement with what we found before. Note that we could choose any other slope triangle, we'd still get the same result.

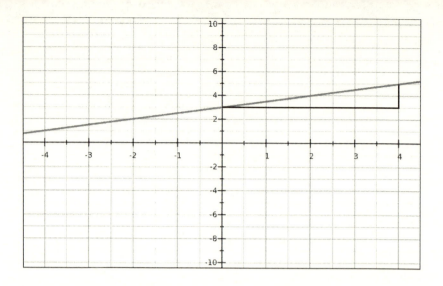

The slope triangle shows that a change in x by Δx = 4 units corresponds to a change in y by Δy = 2 units. Thus:

a = Δy / Δx = 2 / 4 = 0.5

Here's an example that shows the need for drawing in such a well-chosen slope triangle and using the general formula.

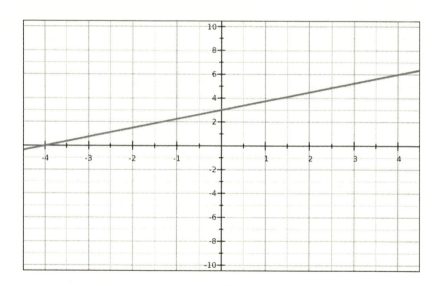

Going only one unit to the right, it's not absolutely clear by how much the graph goes up. Two-thirds of a unit? Three-fourths of a unit? Something else? Mathematics should not be guesswork. So let's find a good slope triangle. It seems that going four units to the right might be just what we need.

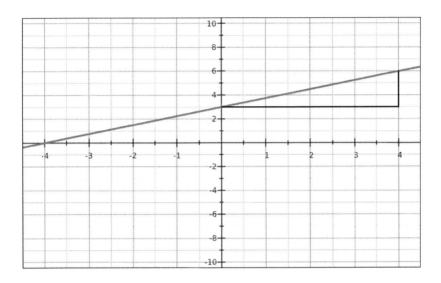

We can see that a change in x by $\Delta x = 4$ units corresponds to a change in y by $\Delta y = 3$ units. The slope is:

$a = \Delta y / \Delta x = 3 / 4 = 0.75$

So our second guess was correct, the graph goes up by three-fourths of a unit when going to the right by one unit. With this formula at hand, we are ready to tackle another important problem often encountered when dealing with linear functions: determining the function that goes through two given points P and Q. Suppose you want to find the linear function that goes through the points P(-3, -2) and Q(2, 4). How could we do this?

S:

Hmmm ... I think we could mark the two points in a coordinate system, draw a straight line through them and then find the y-intercept and slope graphically.

T:

That could work. However, we might run into problems with accuracy. Doing the plot you will see that the y-intercept is somewhere between 1 and 2, but it's hard to tell where exactly the line crosses the y-axis. The graphical approach will give you a good approximation, you might even find the exact solution, but there would be guesswork involved. So let's think of another approach, a purely mathematical one.

S:

Do I need to use the slope formula?

T:

That would be a good idea. Think about the x- and y-changes when going from point P(-3, -2) to point Q(2, 4).

S:

When going from point P to point Q, we go from x = -3 to x = 2, that's five units in total.

T:

That's right, so the change in x is $\Delta x = 5$.

S:

Ah, now I see. From P to Q we also go from y = -2 to y = 4, that's six steps. So the change in y is $\Delta y = 6$. Now that we have the change in x and the change in y, we can use the formula to calculate the slope:

$a = \Delta y / \Delta x = 6 / 5 = 1.2$

T:

Well done! By looking at the changes in x and y, we were able to determine the slope of the linear function that goes through P(-3, -2) and Q(2, 4). The function must thus look like this:

$f(x) = 1.2 \cdot x + b$

With the y-intercept b we still need to find. Any ideas?

S:

I'm not sure how to do this.

T:

Don't worry, I will show you. What do we know about this function? We know that it goes through the point P(-3, -2). In other words: when we use the input x = -3, we want the output to be y = -2. So let's insert x = -3 and demand that the result is y = -2. This leads to a linear equation for b.

$f(x) = 1.2 \cdot x + b$

$f(-3) = 1.2 \cdot (-3) + b = -2$

Dropping the f(-3):

$1.2 \cdot (-3) + b = -2$

Let's solve this!

$1.2 \cdot (-3) + b = -2$

$-3.6 + b = -2 \quad / + 3.6$

$b = -2 + 3.6 = 1.6$

And that's already the solution! We found a = 1.2 and b = 1.6, so the function that goes through P(-3, -2) and Q(2, 4) is:

f(x) = 1.2·x + 1.6

Note that had we used the other point to determine b, we would have gotten the same result. Our function goes through Q(2, 4), meaning that when we insert x = 2, we want the result to be y = 4. The corresponding equation is:

f(x) = 1.2·x + b

f(2) = 1.2·2 + b = 4

Dropping the f(2) and solving:

1.2·2 + b = 4

2.4 + b = 4   / - 2.4

b = 4 - 2.4 = 1.6

So once you've found the slope by looking at the changes in x and y and once you've written down the preliminary result, that is, the function with the known slope a and the undetermined b, you are free to insert any of the two points to find the y-intercept. Let's do this once more. Please find the linear function that goes through the points P(-4, -3) and Q(2, 1).

S:

Okay. When going from P(-4, -3) to Q(2, 1), the variable x changes from x = -4 to x = 2, so the change in x is Δx = 6. At the same time, the variable y changes from y = -3 to y = 1, the total change in y is thus Δy = 4. So the slope must be:

a = Δy / Δx = 4 / 6 = 2 / 3

T:

Very nice. Now write down the preliminary result.

S:

The function must look like this:

f(x) = 2/3·x + b

So now I must insert a point?

T:

That's right. Choose any point. I'd use the point without any negative numbers, this just simplifies the calculation.

S:

Okay. The function goes through Q(2, 1), so when inserting x = 2, the output must be y = 1.

f(x) = 2/3·x + b

f(2) = 2/3·2 + b = 1

The equation and the solution is:

2/3·2 + b = 1

4/3 + b = 1   / - 4/3

b = 1 - 4/3 = -1/3

So this should be the function:

f(x) = 2/3·x - 1/3

Is that correct?

T:

See for yourself. You know the function should go through the points P(-4, -3) and Q(2, 1). So check if for x = -4 you get the output y = -3 and for x = 2 you get the output y = 1.

S:

Fine. Using:

f(x) = 2/3·x - 1/3

I get the following for x = -4:

f(-4) = 2/3·(-4) - 1/3 = -3

And the following for x = 2:

f(2) = 2/3·2 - 1/3 = 1

So it all checks out.

T:

Yes, excellent work. The function f(x) = 2/3·x - 1/3 indeed goes through P(-4, -3) and Q(2, 1). So remember the strategy. To find the linear function that goes through two given points, we first calculate the slope a by looking at the changes in x and y, then write down the preliminary result with the undetermined b and finally insert any of the two points to find the value of b. Then you have the solution.

A quick note on finding the x- and y-changes. When we are given points with non-integer coordinates, this can be a difficult task. For example, suppose I asked you to find the linear function that goes through P(-7/9, 2) and Q(13/9, 3/4). How to find the corresponding changes? The trick is not to think too much and subtract mechanically. This will always lead to the right result. Given the points $P(x_p, y_p)$ and $Q(x_q, y_q)$, the changes in x and y when going from P to Q are:

$\Delta x = x_q - x_p$

$\Delta y = y_q - y_p$

In case of the points P(-7/9, 2) and Q(13/9, 3/4) we get:

$\Delta x = 13/9 - (-7/9) = 20/9$

$\Delta y = 3/4 - 2 = -5/4$

Leading to the slope:

$a = \Delta y / \Delta x$

$a = (-5/4) / (20/9) = (-5/4) \cdot (9/20)$

$a = -45/80 = -9/16$

Note the signs when calculating Δx. The point we are given is P(-7/9, 2) and thus we have the coordinate $x_p = -7/9$ and not $x_p = 7/9$. Accordingly, the change in x is:

$\Delta x = x_q - x_p = 13/9 - (-7/9)$

And not:

$\Delta x = x_q - x_p = 13/9 - 7/9$

As one might be tempted to think. So when given points with non-integer coordinates, feel free to subtract mechanically, but make sure to pay very close attention to the signs. I assure you that nothing will go wrong then.

Okay, there is another topic we should discuss before moving on to a new class of functions: roots (or zeroes). The roots of a function are the values of x at which the value of the function f(x) becomes zero. We can determine the roots by setting up the equation f(x) = 0 and solving for x. This approach works for all functions, though there is no guarantee that the resulting equation can always be solved analytically, that is, by hand. In the geometric interpretation the roots correspond to the values of x at which the graph crosses the x-axis. Let's look at a few examples at how to find the roots of linear functions.

Suppose I want to know at which values of x the value of the function f(x) = 3·x - 2 becomes zero. This was the first linear function we encountered in this lesson, you can see its graph when scrolling back to the beginning. To find the roots, we set up the equation f(x) = 0 and solve for x.

f(x) = 0

3·x - 2 = 0   / + 2

3·x = 2   / : 3

x = 2/3

So the root of f(x) = 3·x - 2 is x = 2/3. At this value of x, the graph of f(x) crosses the x-axis. The graph confirms this. Now it's your turn. Please find all roots of the function f(x) = 2·x + 1, the second linear function encountered in this lesson.

S:

Will do.

f(x) = 0

2·x + 1 = 0   / -1

2·x = -1   / : 2

x = -1/2

So the root is x = -1/2. Is that correct?

T:

See for yourself. Just insert x = -1/2 and see if the function indeed returns the output y = 0.

S:

I'll try that.

f(x) = 2·x + 1

f(-1/2) = 2·(-1/2) + 1 = 0

Yes, at x = -1/2 the value of the function is y = 0. So x = -1/2 is the root of the function.

T:

Great job!

S:

You mentioned that sometimes the equation cannot be solved. Can you show me an example?

T:

Well, for linear functions there is no such equation. The equation f(x) = 0 can always be solved when f(x) is a linear function. The same is true when f(x) is a quadratic or cubic function, though the latter requires tremendous work. But for more complicated functions the equation f(x) = 0 might not allow us to bring it to the form x = ... So when I stated that sometimes f(x) = 0 cannot be solved, that is, we cannot find the roots of the function analytically, I was referring to more complicated functions.

S:

Okay, understood.

T:

Now let's go to the final section of this lesson. I would like to have a look at a real-life application. There are many cases in which two physical quantities are linked in a linear fashion, that is, via some linear function. One example is the dependence of the boiling point on altitude. The higher up you go in the atmosphere, the smaller the temperature at which water begins to boil. The relationship between the two quantities is approximately linear:

$f(x) = a \cdot x + b$

With x being the altitude in meters and f(x) the boiling point in degrees Celsius. Here's some useful data: the elevation of Las Vegas, Nevada, is roughly x = 660 m. Water in Las Vegas, begins to boil at about y = 97.82 °C. In Leadville, Colorado, with its elevation of x = 3100 m the highest city in the US, water already boils at a temperature of y = 89.77 °C. Please do the following:

a) Find the linear function $f(x) = a \cdot x + b$ that goes through the given data points.

b) Calculate at what temperature water starts to boil on the peak of Mount Everest. It lies 8848 m above the sea level.

c) Calculate at which altitude water begins to boil at 80 °C.

S:

That sounds like a lot of work, but I will try. So I basically have to find the linear function that goes through the points P(660, 97.82) and Q(3100, 89.77). I'll start by computing the slope. The formula is:

$a = \Delta y / \Delta x$

The change in y is:

$\Delta y = 89.77 - 97.82 = -8.05$

The change in x is:

Δx = 3100 - 660 = 2440

So the slope is:

a = Δy / Δx = -8.05 / 2440 ≈ -0.0033

Is that good?

T:

Looks fine, go on.

S:

Here's the preliminary function:

f(x) = -0.0033·x + b

I'll now insert one of the two points to find the value of b. I'll try point Q(3100, 89.77), so when inserting x = 3100 the value of the function should be y = 89.77.

f(3100) = -0.0033·3100 + b = 89.77

The equation is:

-0.0033·3100 + b = 89.77

-10.23 + b = 89.77   / + 10.23

b = 89.77 + 10.23 = 100

The function is:

f(x) = -0.0033·x + 100

T:

Very nice. Yes, this is the function that allows you to insert the altitude x in meters and spits out the boiling point y in degrees Celsius. Now try part b) of the exercise. That should be easy.

S:

I think here I just have to use the input x = 8848:

f(8848) = -0.0033 · 8848 + 100 ≈ 70.8

So according to the function, water will boil at 70.8 °C on the peak of Mount Everest. Is this really true?

T:

Indeed. Wikipedia says 71 °C, close enough. Great job, onto part c) of the exercise.

S:

I'm not sure how to do this one. Can I just insert x = 80?

T:

No, that doesn't work. This way you calculate the boiling point y of water at the altitude x = 80 m. But that's not what we want to know. We want to find out at which altitude x the boiling point is y = 80 °C. So here we are given the output y and seek the corresponding input x.

S:

Oh yes. What about this: I set up the equation f(x) = 80 and solve for x. Would that work?

T:

That sounds a lot better.

S:

Okay. Here it is:

f(x) = -0.0033·x + 100 = 80

-0.0033·x + 100 = 80   / - 100

-0.0033·x = -20   / : (-0.0033)

x = -20 / (-0.0033) ≈ 6060

So water begins to boil at 80 degrees Celsius at the altitude 6060 meters. I'll check if this is correct:

f(6060) = -0.0033·6060 + 100 ≈ 80

That seems to work.

T:

Great job! With this I would like to conclude lesson two. We learned that the general form of a linear function is:

f(x) = a·x + b

The constant b is the y-intercept, that is, the value of y at which the graph of the function crosses the y-axis. Constant a is called the slope. Note that some textbooks use the letter "m" instead of "a", but it makes no difference. The slope tells us by how much the graph moves up or down when going one unit to the right. There is a neat formula for calculating it:

$a = \Delta y / \Delta x$

In words: the slope is equal to the change in y divided by the change in x. We can use this formula when it is not clear from the plot by how much the graph goes up or down when going one unit to the right or when we are only given two random points of a linear function.

To determine the linear function that goes through two given points $P(x_p, y_p)$ and $Q(x_q, y_q)$, we first compute the changes in x- and y-direction by subtracting the respective coordinates:

$\Delta x = x_q - x_p$

$\Delta y = y_q - y_p$

Pay close attention to the signs here. We then can apply the formula $a = \Delta y / \Delta x$ to find the slope. With this done, we can write down the preliminary function $f(x) = a \cdot x + b$ with the still undetermined b. To find the value of b, we simply insert one of the two points.

We also learned how to compute the roots of linear functions. We simply set up the equation $f(x) = 0$ and solve for x. The solution tells us at which value of x the line crosses the x-axis. So this is what you should know at this point. Now you can relax, you've earned it, but make sure to come back for lesson three. There's still much to learn about functions.

Lesson III

- **Quadratic Functions**

- **Constants**

- **Vertex Point**

- **Roots**

- **Factored Form**

T:

It's you again! Glad to see that you're back for more. Today I want to discuss quadratic functions, a very common class of functions. We'll start again with a couple of graphs and try to find some sort of pattern. Ready? Here we go.

$f(x) = 2 \cdot x^2 - 1$

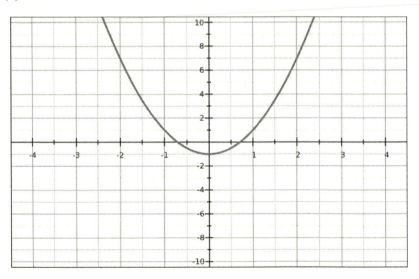

$f(x) = 0.5 \cdot x^2 + 3$

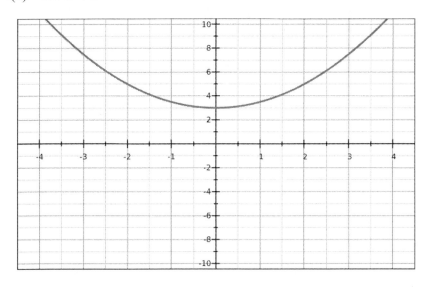

$f(x) = -1.5 \cdot x^2 + 7$

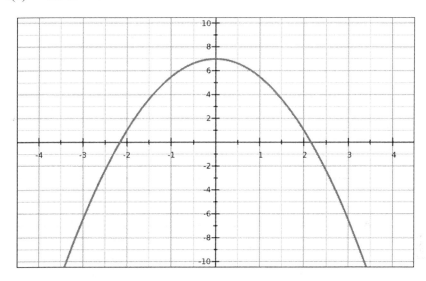

$f(x) = 2 \cdot x^2 - 4 \cdot x - 5$

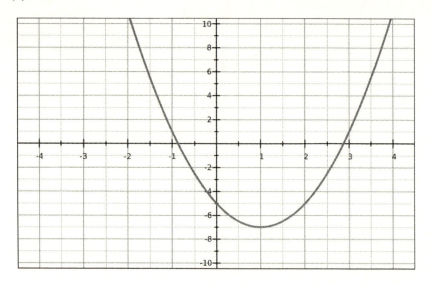

$f(x) = 2 \cdot x^2 + 4 \cdot x - 5$

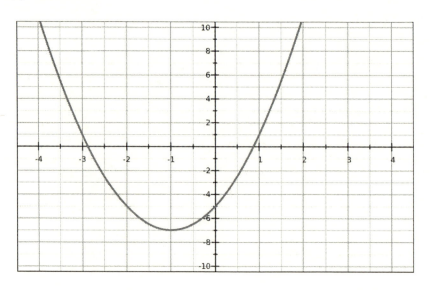

S:

That's a lot of graphs!

T:

It is, isn't it? What do you see? I'll get you started. Quadratic functions have the general form $f(x) = a \cdot x^2 + b \cdot x + c$, with three constants a, b and c. So it's basically a linear equation with an additional quadratic term. The graph of quadratic functions is always a parabola. Now take a look at how the value of constant c influences the graph.

S:

Yes, I see that. Constant c is where the parabola crosses the y-axis. In the first example we have $f(x) = 2 \cdot x^2 - 1$ and the graph crosses the y-axis at $y = -1$. In the second example the function is $f(x) = 0.5 \cdot x^2 + 3$ and it crosses the y-axis at $y = 3$.

T:

Exactly. So we can conclude that constant c is the good old y-intercept. Now have a look at constant b.

S:

The constant b is missing in the first three examples.

T:

Yes, in the first three examples we have a quadratic function with $b = 0$. But I've chosen the last two examples so you might see the influence of constant b on the graph.

S:

Okay, it seems to me that when b = 0, the parabola's axis of symmetry is the y-axis, for b ≠ 0 it isn't.

T:

That's correct. And I would like to leave it at that for now regarding constant b. What about constant a?

S:

Hmm ... in all cases except example number three the constant a is positive and in all cases except example number three the parabola opens upwards. So I think that when a is positive, the parabola opens upwards and when a is negative, the parabola opens downwards. In addition to that, I believe that constant a is also somehow related to how "narrow" the parabola is.

T:

Fantastic observations. All of the above is correct. Constant a determines two things: how the parabola opens and how narrow it is. When a > 0, the parabola opens upwards and when a < 0, it opens downwards. We don't need to consider the case a = 0 since this would bring us back to a linear function. A quadratic function with a = 0 doesn't exist. As for your other observation, it's true that when the magnitude of constant a is large, that is, if it is a large number, positive or negative, the parabola is quite narrow, while for values of a close to zero, it is rather wide. You can see the latter in the following graph.

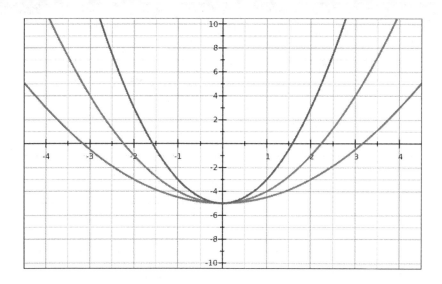

The blue line shows the graph of f(x) = 0.5·x² - 5, the red line the graph of f(x) = x² - 5 and the green line the graph of f(x) = 2·x² - 5. You can see that as the value of a increases, the parabola gets narrower. Now let's put all of the observations together. Here's another graph for you to consider. Please tell me why this graph makes sense given the function.

f(x) = 3·x² + 6·x - 2

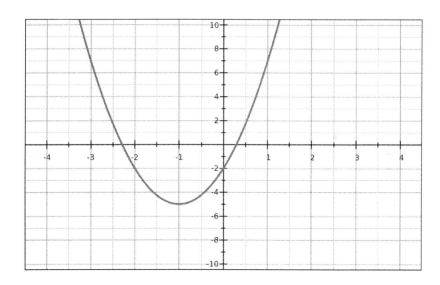

S:

I'll try. The constant c has the value c = -2. Accordingly, the graph should cross the y-axis at y = -2, which it does. Constant b is not equal to zero, meaning that the parabola's axis of symmetry should not be on the y-axis. This is also true. Finally, constant a is greater than zero, so the graph must open upwards. Seems like it all checks out.

T:

Great job! Now please write down the quadratic function corresponding to a parabola that a) crosses the y-axis at y = 5, b) opens downwards, c) is rather wide and d) has its axis of symmetry on the y-axis.

S:

Since the parabola is supposed to cross the y-axis at y = 5, we must have c = 5. It should open downwards, so a must be smaller than zero. Since we want a wide parabola, a should also be close to zero. So how about a = -0.1?

T:

Seems fine, go on.

S:

Okay, finally the parabola's axis of symmetry is supposed to be on the y-axis, this implies b = 0. So the function is:

$f(x) = -0.1 \cdot x^2 + 5$

T:

Correct, the above function satisfies all the conditions. Seems to me that you understood the influence of the constants on the graph. Well done! Then let's move on. I would like to continue with the vertex, that is, the minimum or maximum of the parabola. We can calculate the location of the vertex from the constants using the following formula.

$$x_v = -b / (2 \cdot a)$$

So to find the x-coordinate of the vertex, we divide constant b, with the sign reversed, by two times the constant a. Where does this formula come from? Understanding this requires some heavy algebra, so I prefer to show it to you at the end of the lesson if you are still interested. For now, let's just accept the formula as a fact.

To find the y-coordinate of the vertex, we do what we always do when trying to find the output y corresponding to a certain input x. We throw the x_v that we calculated using the above formula into the function and get the y-coordinate y_v in return. In short: $y_v = f(x_v)$. Then we know at which point $P(x_v, y_v)$ the vertex lies. Here's a quick example. Let's go back to:

$$f(x) = 3 \cdot x^2 + 6 \cdot x - 2$$

From the graph we can see that the vertex of the parabola should be the point P(-1, -5). I want to confirm this analytically. In this case we have b = 6 and a = 3. Inserting this into the formula for the x-coordinate of the vertex leads to:

$$x_v = -b / (2 \cdot a)$$

$$x_v = -6 / (2 \cdot 3) = -6 / 6 = -1$$

So the x-coordinate of the vertex is $x_v = -1$. So far, so good! To find the y-coordinate of the vertex we simply insert x = -1 into the quadratic function. We get:

$$f(x) = 3 \cdot x^2 + 6 \cdot x - 2$$

$$f(-1) = 3 \cdot (-1)^2 + 6 \cdot (-1) - 2$$

$f(-1) = 3 - 6 - 2 = -5$

Since $(-1)^2 = 1$. This shows that the y-coordinate of the vertex is $y_v = -5$. The calculation thus confirms that the point $P(-1, -5)$ is the vertex of the parabola. That wasn't so bad, right? Now please calculate the location of the vertex of the function:

$f(x) = -x^2 + 4 \cdot x + 2$

After that, go to graphsketch.com and confirm your calculation by having the website generate the plot of said function.

S:

Okay, I need to find the values of a and b and use the formula $x_v = -b/(2 \cdot a)$. Clearly $b = 4$, but what about a?

T:

Note that $x^2 = 1 \cdot x^2$ and thus $-x^2 = -1 \cdot x^2$. Does that help?

S:

Ah, of course. So we have $b = 4$ and $a = -1$. So:

$x_v = -b/(2 \cdot a)$

$x_v = -4/(2 \cdot (-1)) = -4/(-2) = 2$

The x-coordinate of the vertex should be $x_v = 2$. I'll insert this into the function to find the corresponding value of y.

$f(x) = -x^2 + 4 \cdot x + 2$

$f(2) = -2^2 + 4 \cdot 2 + 2$

$f(2) = -4 + 8 + 2 = 6$

This implies that the y-coordinate of the vertex is $y_v = 6$, so the vertex of the parabola is the point P(2, 6). The plot of the function confirms this:

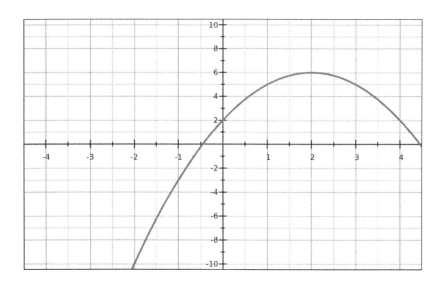

T:

Looking good! Here are a few general conclusions from the formula $x_v = -b / (2 \cdot a)$. We can see that for $b = 0$ we get $x_v = 0$, meaning that for quadratic functions with $b = 0$, the vertex, and thus the axis of symmetry, is on the y-axis. This is in line with what we observed earlier.

Note that if a and b have the same signs, the ratio b / a is always positive. For example, for $a = 2$ and $b = 8$ we have $b / a = 4 > 0$ and for $a = -2$ and $b = -8$ we also have $b / a = 4 > 0$. This implies that if a and b have the same signs, the x-coordinate of the vertex is always negative, that is, the vertex and with it the axis of symmetry lies to the left of the y-axis. Using the same argument we can show that if a and b do not have the same signs, the vertex and the axis of symmetry must be to the right of the y-axis. Revisit all the graphs to confirm this.

Okay, we now know how to find the vertex of any given quadratic function. We simply extract the values of constants a and b, apply the formula $x_v = -b / (2 \cdot a)$ and finally insert what we found for x_v into the function to get y_v. But what about the other way around? Suppose we are given the location of the vertex and we want to find a quadratic function that goes through it. How can we do this?

Again, finding the solution requires quite a bit of algebra, we won't do that right now. But rest assured, I'll show it to you at the end of the lesson. For now, just believe me when I say that we can setup a quadratic function that goes through a given vertex $P(x_v, y_v)$ using the formula:

$$f(x) = a \cdot (x - x_v)^2 + y_v$$

With the still undetermined constant a that will either be given as well or must be determined from some other condition such as an additional point the graph should go through. Let me show you an example. Suppose I want to find a quadratic function having the vertex $P(1, -6)$. We also demand that $a = 2$. Inserting this into the above formula we get:

$$f(x) = a \cdot (x - x_v)^2 + y_v$$

$$f(x) = 2 \cdot (x - 1)^2 - 6$$

We could already stop here, but generally we want to bring it to the standard form $f(x) = a \cdot x^2 + b \cdot x + c$. Let's do that now. I hope you remember the binomial formulas!

S:

Oh, it's been a while.

T:

Okay, no problem. Then we'll have a quick look at the first two binomial formulas before going back to the example.

$$(x + y)^2 = x^2 + 2 \cdot x \cdot y + y^2$$

$$(x - y)^2 = x^2 - 2 \cdot x \cdot y + y^2$$

As you can see, we can use these formulas to expand squared brackets, which is just what we need to do to bring the function $f(x) = 2 \cdot (x - 1)^2 - 6$ to the standard form $f(x) = a \cdot x^2 + b \cdot x + c$. Here are a few examples of how to apply the two binomial formulas. It's really not that difficult.

$(x + 5)^2 = x^2 + 2 \cdot x \cdot 5 + 5^2 = x^2 + 10 \cdot x + 25$

$(x + 2)^2 = x^2 + 2 \cdot x \cdot 2 + 2^2 = x^2 + 4 \cdot x + 4$

$(x - 3)^2 = x^2 - 2 \cdot x \cdot 3 + 3^2 = x^2 - 6 \cdot x + 9$

$(x - 10)^2 = x^2 - 2 \cdot x \cdot 10 + 10^2 = x^2 - 20 \cdot x + 100$

So to expand $(x + y)^2$ you simply square x, add to that two times x times y and finally add y squared to this. If instead of the plus- you have the minus-sign in the bracket, you square x, subtract from that two times x times y and finally add y squared. That's really all you need to know.

Let's return to the example. We want to bring the following function, which we know represents a parabola having its vertex at P(1, -6), to the standard form:

$f(x) = 2 \cdot (x - 1)^2 - 6$

Using the second binomial formula, we find that:

$(x - 1)^2 = x^2 - 2 \cdot x \cdot 1 + 1^2 = x^2 - 2 \cdot x + 1$

Going back to the function:

$f(x) = 2 \cdot (x - 1)^2 - 6$

$f(x) = 2 \cdot (x^2 - 2 \cdot x + 1) - 6$

$f(x) = 2 \cdot x^2 - 4 \cdot x + 2 - 6$

And finally:

$f(x) = 2 \cdot x^2 - 4 \cdot x - 4$

The standard form of the quadratic function with the vertex P(1, -6) and a = 2. From the image below you can see that our approach really works.

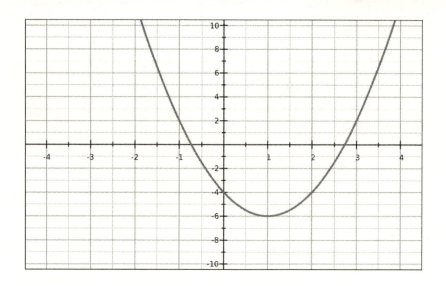

It's your turn! Please find the quadratic function having the vertex P(-2, 8) and constant a = -3.

S:

Hmmm ... Okay. I think I need to use:

$f(x) = a \cdot (x - x_v)^2 + y_v$

T:

That's right.

S:

Inserting $P(x_v, y_v) = P(-2, 8)$ and a = -3 leads to:

$f(x) = -3 \cdot (x - 2)^2 + 8$

T:

Oh, careful with the signs here! We have $x_v = -2$ and thus $x - x_v = x - (-2) = x + 2$. Please correct that.

S:

Of course. So we have:

$f(x) = -3 \cdot (x + 2)^2 + 8$

T:

Yes, this function has the vertex P(-2, 8) and constant $a = -3$. Now bring it to the standard form, please.

S:

According to the first binomial formula:

$(x + y)^2 = x^2 + 2 \cdot x \cdot y + y^2$

We have:

$(x + 2)^2 = x^2 + 2 \cdot x \cdot 2 + 2^2 = x^2 + 4 \cdot x + 4$

Inserting this into the function:

$f(x) = -3 \cdot (x + 2)^2 + 8$

$f(x) = -3 \cdot (x^2 + 4 \cdot x + 4) + 8$

$f(x) = -3 \cdot x^2 - 12 \cdot x - 12 + 8$

$f(x) = -3 \cdot x^2 - 12 \cdot x - 4$

Is that right?

T:

Well, let's ask the plot, shall we?

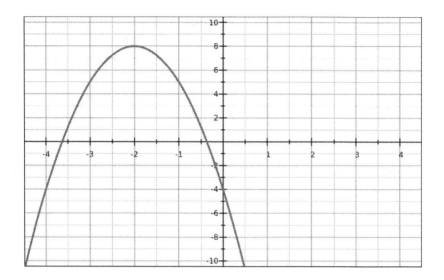

Looks like it all checks out, the vertex is at P(-2, 8). That was excellent work!

S:

Quick question: why do we have to bring it to the standard form? Isn't the answer $f(x) = -3 \cdot (x + 2)^2 + 8$ good enough?

T:

It actually is. When somebody asks you to find the quadratic function having the vertex P(-2, 8) and constant a = -3 and your reply is $f(x) = -3 \cdot (x + 2)^2 + 8$, then there is nothing wrong with that answer. But I demanded to see the corresponding standard form because it's generally easier to work with a function in standard form. Sometimes we have to do more things with the function, such as finding the roots of it, and in this case it's very helpful to have the standard form.

Okay, now let's turn it up a bit. Up to now were given the vertex as well as the constant a. However, the latter is rather unrealistic. It would make much more sense to ask for a quadratic function with a given vertex P and an additional point Q the graph should go through. Let's try this out.

Suppose we want to find the quadratic function that has the vertex P(-3, -9) and also goes through Q(-1, 8). Locate the points in the coordinate system and visualize how the parabola should look like. How to approach this? Well, our best guess is to again start with the formula:

$$f(x) = a \cdot (x - x_v)^2 + y_v$$

Inserting the vertex P(-3, -9) we get:

$$f(x) = a \cdot (x + 3)^2 - 9$$

Do you have an idea how we can determine the value of a given that the graph should also go through Q(-1, 8)?

S:

Hmmm ... The point Q(-1, 8) says that for x = -1 the value of the function should be y = 8. Can we make some sort of equation from that?

T:

Indeed we can. Set up f(-1) = 8.

S:

I'll try. Inserting x = -1 leads to:

$$f(-1) = a \cdot (-1 + 3)^2 - 9$$

$$f(-1) = a \cdot 2^2 - 9$$

$$f(-1) = 4 \cdot a - 9$$

Now this should be equal to 8, right?

T:

Correct.

S:

So I think the equation is:

4·a - 9 = 8

T:

Very nice, a linear equation for constant a. That shouldn't be too hard to solve. Just bring it to the form a = ...

S:

Okay.

4·a - 9 = 8   / + 9

4·a = 17   / : 4

a = 17 / 4 = 4.25

Inserting this into:

f(x) = a·(x + 3)² - 9

Leads to:

f(x) = 4.25·(x + 3)² - 9

T:

Great job! That's the function that has its vertex at P(-3, -9) and also goes through Q(-1, 8). Check out the graph.

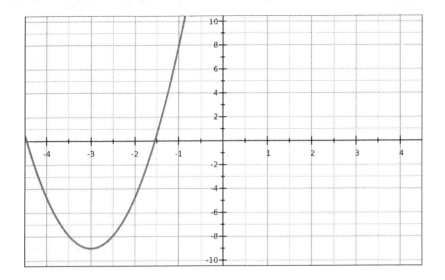

Since the problem is now solved, I'll leave it up to you whether you want to bring it to the standard form or not.

S:

I want to try it out. The function is:

$f(x) = 4.25 \cdot (x + 3)^2 - 9$

According to the binomial formulas:

$(x + 3)^2 = x^2 + 2 \cdot x \cdot 3 + 3^2 = x^2 + 6 \cdot x + 9$

So:

$f(x) = 4.25 \cdot (x + 3)^2 - 9$

$f(x) = 4.25 \cdot (x^2 + 6 \cdot x + 9) - 9$

$f(x) = 4.25 \cdot x^2 + 25.5 \cdot x + 38.25 - 9$

$f(x) = 4.25 \cdot x^2 + 25.5 \cdot x + 29.25$

T:

Excellent! Let's do one more exercise of this type before leaving the topic "vertex" and moving on to the roots of a quadratic function. Please find the quadratic function that has its vertex at P(1, -3) and goes through the origin Q(0, 0).

S:

The formula we need is:

$f(x) = a \cdot (x - x_v)^2 + y_v$

With the vertex at P(1, -3) we have:

$f(x) = a \cdot (x - 1)^2 - 3$

The function also goes through the origin, so for x = 0 the output of the function must be y = 0.

$f(0) = a \cdot (0 - 1)^2 - 3 = 0$

$a \cdot (0 - 1)^2 - 3 = 0$

$a \cdot (-1)^2 - 3 = 0$

$a - 3 = 0 \quad / + 3$

$a = 3$

So the function we were looking for is:

$f(x) = 3 \cdot (x - 1)^2 - 3$

T:

Yes, that's the one! And here's the graph of f(x).

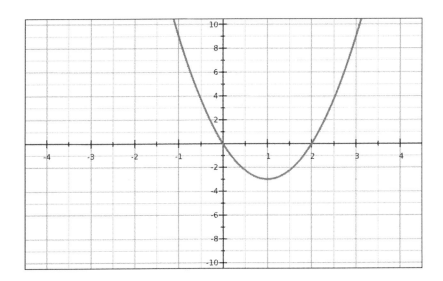

Okay, fantastic. Then let's move on to the roots of quadratic functions. As you might recall, the roots of a function are the values of x at which the output y is zero, that is, the values of x at which the graph of the function crosses the x-axis. We can find the roots by setting up the equation f(x) = 0 and solving for x. So here's what we need to solve:

$a \cdot x^2 + b \cdot x + c = 0$

How to do this? Bringing it to the form x = ... using straight-forward algebra is possible, but quite difficult. That's why we will use a general formula to find the solution(s). Now don't be shocked by the formula, it really looks worse than it actually is. Plus, the following examples will make it crystal-clear how to apply it. Without further ado, here's the grand formula we can use to solve the above quadratic equation:

$x_{1,2} = (-b \pm \text{sqrt}(b^2 - 4 \cdot a \cdot c)) / (2 \cdot a)$

This is called the quadratic formula. Let's see how it works. Suppose we want to find out where the graph of the following function crosses the x-axis.

$f(x) = 2 \cdot x^2 + 4 \cdot x - 6$

Here's the graph of $f(x)$:

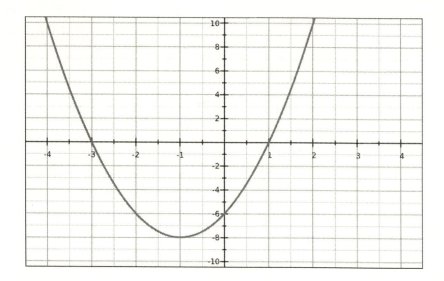

According to this, the roots should be $x = -3$ and $x = 1$. Let's confirm that. From the function we can see that the constants are $a = 2$, $b = 4$ and $c = -6$. Note that the signs are included here. Now we throw all of the constants into the quadratic formula and simplify.

$x_{1,2} = (-b \pm sqrt(b^2 - 4 \cdot a \cdot c)) / (2 \cdot a)$

$x_{1,2} = (-4 \pm sqrt(4^2 - 4 \cdot 2 \cdot (-6))) / (2 \cdot 2)$

$x_{1,2} = (-4 \pm sqrt(16 + 48)) / 4$

$x_{1,2} = (-4 \pm sqrt(64)) / 4$

$x_{1,2} = (-4 \pm 8) / 4$

So what to do now? We calculate the first root x_1 by choosing the plus-sign. This leads to:

$x_1 = (-4 + 8) / 4 = 1$

So the first root is $x_1 = 1$. To compute the second root we choose the minus-sign.

$x_2 = (-4 - 8) / 4 = -3$

The second root is thus $x_2 = -3$. Both results are in line with what we expected. That wasn't so bad, was it? So to find the roots of a quadratic function, you should first set up the equation $f(x) = 0$, note the values of the constants a, b and c and insert these values into the quadratic formula. Simplify, then choose the plus-sign for the first root x_1 and the minus-sign for the second root x_2. That's all.

Here's a tip: pay very, very close attention to the signs! If you got the wrong solution, I'll bet you a thousand dollars that somewhere there's a problem with the signs. I've been teaching algebra for many years and I can tell you from experience that when it comes to solving quadratic equations, incorrect signs account for 90 % of the mistakes. So spare yourself the frustration and handle the signs with care.

S:

Will do.

T:

Okay, here's a job for you. Please find the roots of the following quadratic function.

$f(x) = -3 \cdot x^2 + 3 \cdot x + 6$

S:

So here we have to solve:

$-3 \cdot x^2 + 3 \cdot x + 6 = 0$

The constants are a = -3, b = 3 and c = 6. I'll insert them into the quadratic formula and simplify:

$x_{1,2} = (-b \pm \text{sqrt}(b^2 - 4 \cdot a \cdot c)) / (2 \cdot a)$

$x_{1,2} = (-3 \pm \text{sqrt}(3^2 - 4 \cdot (-3) \cdot 6)) / (2 \cdot (-3))$

$x_{1,2} = (-3 \pm \text{sqrt}(9 + 72)) / (-6)$

$x_{1,2} = (-3 \pm \text{sqrt}(81)) / (-6)$

$x_{1,2} = (-3 \pm 9) / (-6)$

Choosing the plus-sign for the first solution:

$x_1 = (-3 + 9) / (-6) = -1$

The first solution is thus $x_1 = -1$. Now the same, but with the minus-sign for the second solution:

$x_2 = (-3 - 9) / (-6) = 2$

So $x_2 = 2$.

T:

Spectacular! Yes, $x_1 = -1$ and $x_2 = 2$ are the roots of the quadratic function $f(x) = -3 \cdot x^2 + 3 \cdot x + 6$. You can also see this from the graph of the function. Note how the graph crosses the x-axis at $x_1 = -1$ and $x_2 = 2$.

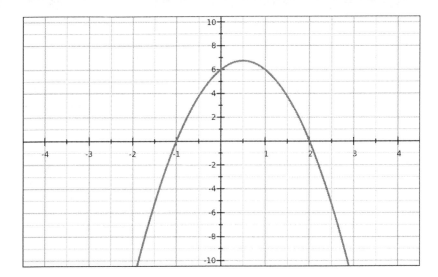

S:

Cool.

T:

Cool indeed. Note that you might get a surprise while trying to find the roots of a quadratic function. Here's what I mean. Please find the roots of:

$f(x) = x^2 - 2 \cdot x + 1$

S:

Setting $f(x) = 0$ leads to the equation:

$x^2 - 2 \cdot x + 1 = 0$

We have a = 1, b = -2 and c = 1. So:

$$x_{1,2} = (\,-b \pm \sqrt{b^2 - 4 \cdot a \cdot c}\,) / (2 \cdot a)$$

$$x_{1,2} = (\,-(-2) \pm \sqrt{(-2)^2 - 4 \cdot 1 \cdot 1}\,) / (2 \cdot 1)$$

$$x_{1,2} = (\,2 \pm \sqrt{4 - 4}\,) / 2$$

$$x_{1,2} = (\,2 \pm \sqrt{0}\,) / 2$$

$$x_{1,2} = (\,2 \pm 0\,) / 2$$

Choosing the plus-sign for the first root:

$$x_1 = (\,2 + 0\,) / 2 = 1$$

And the minus-sign for the second root:

$$x_2 = (\,2 - 0\,) / 2 = 1$$

Hmmm ... Seems like the two roots are the same. How can that be? What does that mean?

T:

This means that the function $f(x) = x^2 - 2 \cdot x + 1$ has only one root, that is, its graph crosses the y-axis at only one value of x and this value is x = 1. The graph agrees with this:

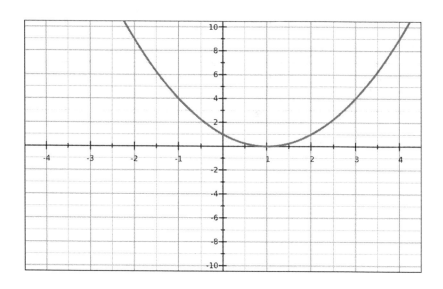

S:

Oh, I see. The graph only touches the x-axis, so there's only one root. That makes sense.

T:

Yes, and note how the quadratic formula automatically takes care of this. By producing the zero under the square root sign, the formula ensures that we only get one solution. Neat, huh? So don't let this special case surprise you. There's another surprise you might encounter. Please find the roots of:

$f(x) = 3 \cdot x^2 + 6 \cdot x + 5$

S:

No problem. The equation we must solve is:

$3 \cdot x^2 + 6 \cdot x + 5 = 0$

The constants are a = 3, b = 6 and c = 5. Hence:

$x_{1,2} = (-b \pm sqrt(b^2 - 4 \cdot a \cdot c)) / (2 \cdot a)$

$x_{1,2} = (-6 \pm sqrt(6^2 - 4 \cdot 3 \cdot 5)) / (2 \cdot 3)$

$x_{1,2} = (-6 \pm sqrt(36 - 60)) / 6$

$x_{1,2} = (-6 \pm sqrt(-24)) / 6$

Oh, my calculator spits out an error message. Of course, there's the square root of a negative number.

T:

Yes, the square root of a negative number is not a real number! We must thus conclude that the function f(x) = 3·x² + 6·x + 5 has no roots at all, that is, the graph never crosses the x-axis. Again we consult the graph to make sure our result is correct.

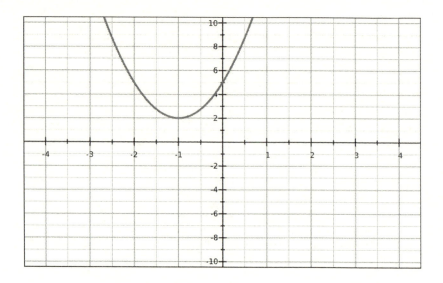

So keep in mind that while for linear functions there is always exactly one root, in case of quadratic functions you can have two, one or no roots. This makes sense considering that a parabola can cross the x-axis twice, once or not at all. Which of the cases applies can be seen from the number you get under the square root sign in the quadratic formula. If this number is positive, there will be two distinct roots, if it is zero, there will be one root and for a negative number you can stop the calculation and note that there is no real-valued root.

Once you found the roots of a quadratic function, you can bring the function to a form that is quite useful sometimes. As you know, the standard form is:

f(x) = a·x² + b·x + c

Given the roots x_1 and x_2, we can also write:

$$f(x) = a \cdot (x - x_1) \cdot (x - x_2)$$

This is called the factored form. For example, a moment ago we found that the roots of the function:

$$f(x) = 2 \cdot x^2 + 4 \cdot x - 6$$

Are $x_1 = 1$ and $x_2 = -3$. The corresponding factored form of the function is thus given by:

$$f(x) = 2 \cdot (x - 1) \cdot (x + 3)$$

By expanding we can see that the two expressions are indeed equal. Remember that when multiplying brackets, we need to multiply each term of bracket one with each term of bracket two. You can do that using the FOIL method: first, outer, inner last, that is, multiply the first terms of the brackets, the outer terms, the inner terms and finally the last terms.

$$f(x) = 2 \cdot (x - 1) \cdot (x + 3)$$

$$f(x) = 2 \cdot (x^2 + 3 \cdot x - 1 \cdot x - 3)$$

$$f(x) = 2 \cdot (x^2 + 2 \cdot x - 3)$$

$$f(x) = 2 \cdot x^2 + 4 \cdot x - 6$$

So it checks out. Please bring:

$$f(x) = x^2 - 5 \cdot x + 4$$

To the factored form.

S:

Do I need to calculate the roots first?

T:

Of course!

S:

Setting f(x) = 0 leads to:

$x^2 - 5 \cdot x + 4 = 0$

We have a = 1, b = -5 and c = 4.

$x_{1,2} = (-b \pm \sqrt{b^2 - 4 \cdot a \cdot c}) / (2 \cdot a)$

$x_{1,2} = (-(-5) \pm \sqrt{(-5)^2 - 4 \cdot 1 \cdot 4}) / (2 \cdot 1)$

$x_{1,2} = (5 \pm \sqrt{25 - 16}) / 2$

$x_{1,2} = (5 \pm \sqrt{9}) / 2$

$x_{1,2} = (5 \pm 3) / 2$

So we have:

$x_1 = (5 + 3) / 2 = 4$

And:

$x_2 = (5 - 3) / 2 = 1$

So the factored form of:

$f(x) = x^2 - 5 \cdot x + 4$

Must be:

$f(x) = a \cdot (x - x_1) \cdot (x - x_2)$

$f(x) = (x - 1) \cdot (x - 4)$

T:

That's right, great job. Note that we can use the factored form to easily set up a quadratic function that has two given roots. For example, suppose I want to construct a quadratic function having the roots $x_1 = -4$ and $x_2 = 2$. I can write:

$f(x) = a \cdot (x - x_1) \cdot (x - x_2)$

$f(x) = a \cdot (x + 4) \cdot (x - 2)$

Since there's no other condition to fulfill, I'm free to choose any value for a, so let's set a = 1. This leads to:

$f(x) = (x + 4) \cdot (x - 2)$

Or in the standard form:

$f(x) = x^2 - 2 \cdot x + 4 \cdot x - 8$

$f(x) = x^2 + 2 \cdot x - 8$

This quadratic function has its roots at $x_1 = -4$ and $x_2 = 2$. We can check that by inserting these values of x and hoping that the output will be y = 0 for each.

$f(-4) = (-4)^2 + 2 \cdot (-4) - 8 = 16 - 8 - 8 = 0$

$f(2) = 2^2 + 2 \cdot 2 - 8 = 4 + 4 - 8 = 0$

Yes, that checks out. Now it's your turn. Can you please "design" a quadratic function in standard form that has the roots $x_1 = -2$ and $x_2 = -1$?

S:

I think I can. I start with:

$f(x) = a \cdot (x - x_1) \cdot (x - x_2)$

Inserting $x_1 = -2$ and $x_2 = -1$:

$f(x) = a \cdot (x + 2) \cdot (x + 1)$

I can choose constant a freely? So I'll also use a = 1.

$f(x) = (x + 2) \cdot (x + 1)$

$f(x) = x^2 + 1 \cdot x + 2 \cdot x + 2$

$f(x) = x^2 + 3 \cdot x + 2$

T:

Very nice. Okay, let's make this a bit more difficult. Now I would like to see a quadratic function that has the roots $x_1 = -1.5$ and $x_2 = 0.5$ and goes through the point P(-0.5, 6). Can you find the right function?

S:

Oh, that sounds tough. Well, since we are given the two roots $x_1 = -1.5$ and $x_2 = 0.5$, it might be a good idea to start with:

$f(x) = a \cdot (x - x_1) \cdot (x - x_2)$

Inserting the roots:

$f(x) = a \cdot (x + 1.5) \cdot (x - 0.5)$

Now I guess I have to determine constant a somehow. Hmmm ... The function must go through P(-0.5, 6), so for the input $x = -0.5$ I want the output to be $y = 6$. Do I have to insert $x = -0.5$ into the function and equate this with 6? That could work, I'll try it out.

We have:

$f(x) = a \cdot (x + 1.5) \cdot (x - 0.5)$

So:

$f(-0.5) = a \cdot (-0.5 + 1.5) \cdot (-0.5 - 0.5) = 6$

$a \cdot (-0.5 + 1.5) \cdot (-0.5 - 0.5) = 6$

$a \cdot 1 \cdot (-1) = 6$

$-a = 6 \quad / \cdot (-1)$

$a = -6$

I think the correct function is:

$f(x) = -6 \cdot (x + 1.5) \cdot (x - 0.5)$

T:

We can consult the graph to see if it checks out. Actually, try to confirm mathematically that the above functions satisfies all the conditions before looking at the graph.

S:

Oh, okay. The roots of the function are supposed to be at $x_1 = -1.5$ and $x_2 = 0.5$, so both these inputs should lead to $y = 0$.

$f(-1.5) = -6 \cdot (-1.5 + 1.5) \cdot (-1.5 - 0.5)$

$f(-1.5) = -6 \cdot 0 \cdot (-2) = 0$

And for the second root:

$f(0.5) = -6 \cdot (0.5 + 1.5) \cdot (0.5 - 0.5)$

$f(0.5) = -6 \cdot 2 \cdot 0 = 0$

Yes, the function crosses the x-axis at $x_1 = -1.5$ and $x_2 = 0.5$. The function should also go through P(-0.5, 6).

$f(-0.5) = -6 \cdot (-0.5 + 1.5) \cdot (-0.5 - 0.5)$

$f(-0.5) = -6 \cdot 1 \cdot (-1) = 6$

That also works. So $f(x) = -6 \cdot (x + 1.5) \cdot (x - 0.5)$ is definitely the function we were looking for.

T:

Well done! And here's the graph.

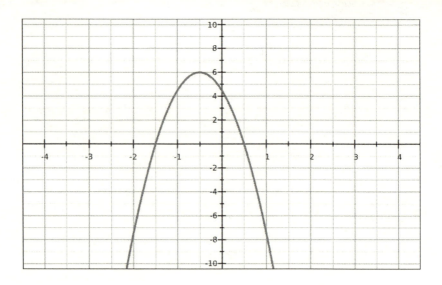

Okay, I think that should be enough material for now. Time to recap. We learned that the standard form of a quadratic function is $f(x) = a \cdot x^2 + b \cdot x + c$, with three constants a, b and c. Constant c is the y-intercept, that is, the value of y at which the parabola crosses the y-axis. Constant b is related to the axis of symmetry. For b = 0, when the linear term is missing, the parabola's axis of symmetry is the y-axis, otherwise it is not. From constant a we can see how the parabola opens. For a > 0, it opens upwards, for a < 0 downwards. That's what you should know about the constants.

All quadratic functions have a vertex, that is, a minimum or maximum. We can calculate the x-coordinate of the vertex using the following formula:

$x_v = -b / (2 \cdot a)$

For the y-coordinate of the vertex we simply throw the x-coordinate x_v into the function: $y_v = f(x_v)$. To set up a quadratic function with a given vertex $P(x_v, y_v)$ we can use:

$f(x) = a \cdot (x - x_v)^2 + y_v$

The constant a is undetermined here. It can be chosen freely if there is no other condition our function must fulfill or must be computed using one additional point. This is not so difficult. Once the vertex and the additional point is inserted, all that remains is a linear equation for a we can easily solve.

At this point you should also know how to calculate the roots of a given quadratic function. You set up the equation f(x) = 0, note the values of the constants a, b and c and insert them into the so-called quadratic formula:

$x_{1,2}$ = (-b ± sqrt(b² - 4·a·c)) / (2·a)

Choose the plus-sign to find the first root x_1 and the minus-sign for the second root x_2. At these values of x the parabola crosses the x-axis. You will only get two distinct roots if the number under the square root sign, or in our case, in the square root operator, is positive. If said number is zero, only one root exists, meaning that the parabola only touches the x-axis. And for a negative number under the square root sign, no real-valued roots exist. Once you know the roots x_1 and x_2, you can also bring the quadratic function to the factored form:

f(x) = a·(x - x_1)·(x - x_2)

Of course you can also use the above formula to design a quadratic function with two given roots x_1 and x_2. In this case constant a is still undetermined. Again, choose it freely if there's no other condition to fulfill, otherwise compute it from an additional point.

That almost concludes lesson three. I hope you enjoyed the learning experience! Make sure to continue your journey into mathematics. If you don't know where to go, check out my "Math Shorts" series, in particular the book "Math Shorts - Exponential and Trigonometric Functions". You should be ready for it and I'm sure you'll learn a lot of interesting things.

But before we go, I still owe you an explanation. I promised to show you where the formula for the vertex comes from. I'll do that now. But beware, it requires some brutal algebra, so the following is meant for those who have finished advanced algebra classes. All others are invited to revisit the following explanation once they have completed such a class. Let's start by rewriting the standard form in a seemingly random manner:

$f(x) = a \cdot x^2 + b \cdot x + c$

$f(x) = a \cdot (x^2 + (b/a) \cdot x) + c$

Clearly I can add a certain number and subtract it again without producing any change. For example: 5 = 5 + 3 - 3. In the same way I can add any expression and subtract it again without altering whatever is there: y = y + z - z. This justifies the following algebraic manipulation:

$f(x) = a \cdot (x^2 + (b/a) \cdot x) + c$

$f(x) = a \cdot (x^2 + (b/a) \cdot x + (b/(2 \cdot a))^2 - (b/(2 \cdot a))^2) + c$

I'll bring one of the terms out of the bracket:

$f(x) = a \cdot (x^2 + (b/a) \cdot x + (b/(2 \cdot a))^2 - (b/(2 \cdot a))^2) + c$

$f(x) = a \cdot (x^2 + (b/a) \cdot x + (b/(2 \cdot a))^2) - a \cdot (b/(2 \cdot a))^2 + c$

$f(x) = a \cdot (x^2 + (b/a) \cdot x + (b/(2 \cdot a))^2) - b^2/(4 \cdot a) + c$

What's the point of all this? Note that according to the first binomial formula, it holds true that:

$(x + b/(2 \cdot a))^2 = x^2 + 2 \cdot x \cdot b/(2 \cdot a) + (b/(2 \cdot a))^2$

$(x + b/(2 \cdot a))^2 = x^2 + (b/a) \cdot x + (b/(2 \cdot a))^2$

This is exactly the expression we have in the bracket. Hence:

$f(x) = a \cdot (x^2 + (b/a) \cdot x + (b/(2 \cdot a))^2) - b^2/(4 \cdot a) + c$

$f(x) = a \cdot (x + b/(2 \cdot a))^2 - b^2/(4 \cdot a) + c$

Now, the square of any number is greater than or equal to zero, for example $(-2)^2 = 4$, $(-1)^2 = 1$, $0^2 = 0$, $1^2 = 1$, $2^2 = 4$, and so on. So for any number or expression y, we have $y^2 \geq 0$. Accordingly:

$(x + b/(2 \cdot a))^2 \geq 0$

No matter what's happening inside the bracket. If a is greater than zero, which we will assume for now, we also have:

$a \cdot (x + b/(2 \cdot a))^2 \geq 0$

Adding $-b^2/(4\cdot a) + c$ to both sides of the inequality, an operation we are allowed to do, leads to:

$a\cdot(x + b/(2\cdot a))^2 - b^2/(4\cdot a) + c \geq -b^2/(4\cdot a) + c$

$f(x) \geq -b^2/(4\cdot a) + c$

So the value of the function is always greater than or equal to $-b^2/(4\cdot a) + c$ given that $a > 0$, that is, in case the parabola opens upwards. So $-b^2/(4\cdot a) + c$ is the minimum value. At which x does the function take on this value? The minimum value is assumed when the bracket $(x + b/(2\cdot a))^2$ equals zero. For all other values of x the bracket will be greater than zero and the value of the function thus greater than the minimum. So for which values of x does the bracket become zero? We can easily find this out:

$(x + b/(2\cdot a))^2 = 0$

$x + b/(2\cdot a) = 0 \quad / - b/(2\cdot a)$

$x = -b/(2\cdot a)$

So the function takes on its minimum value at $x = -b/(2\cdot a)$. Since the minimum is the vertex, we can also say that the x-coordinate of the vertex is $x = -b/(2\cdot a)$. This is what we wanted to show. That was tough! Note that we assumed $a > 0$, but we can use a similar approach to show that the formula $x_v = -b/(2\cdot a)$ also holds true for the case $a < 0$. Also, using some clever manipulation we earlier found that:

$f(x) = a\cdot(x + b/(2\cdot a))^2 - b^2/(4\cdot a) + c$

Given that the x-coordinate of the vertex is $x_v = -b/(2\cdot a)$ and the corresponding y-coordinate $y_v = -b^2/(4\cdot a) + c$, this is because at $x = x_v$ the bracket becomes zero, leaving only the expression $-b^2/(4\cdot a) + c$, so $f(x_v) = y_v = -b^2/(4\cdot a) + c$, we can rewrite this as:

$f(x) = a\cdot(x + b/(2\cdot a))^2 - b^2/(4\cdot a) + c$

$f(x) = a\cdot(x - x_v)^2 + y_v$

The good old vertex form of the quadratic function. So we have also managed to show where this formula comes from. And at this point we'll conclude the final lesson. I hope you had fun and be sure to continue learning. Or as Benjamin Franklin, a brilliant author as well as one of the Founding Fathers of the United States of America, once said:

"An investment in knowledge pays the best interest."

There's no arguing that.

Excerpt from "Business Math Basics" by Metin Bektas

1. Interest

Interest plays a fundamental role in economics. If you loan money to someone, you get it back with interest. And if you borrow money, you pay it back with interest. For most of us this is common sense, but despite that, let's take a quick look at why we give or are given interest at all.

One obvious reason is inflation. If you loan 1000 $ to someone for a period of five years, it will be devalued by the time you get it back. So charging an interest rate at least in the order of the inflation rate seems fair. After all, why should you lose money for providing a favor?

Another reason is the risk of default. You can never be 100 % sure that you get the money you loaned back, even if the borrower is a very good friend, a bank or a state. All of them could make bad investments and go broke. So it seems logical to couple the interest rate to the perceived risk. The higher the risk, the more can be charged. To see a calculation on that, make sure to check out chapter 5.3. (Expected Value) of this book.

Another point is what economists call "opportunity costs". You can't make any (possibly very lucrative) investments with the money you loaned or use it to buy something. So the process of loaning money costs you opportunities. This inconvenience is compensated by charging interest.

There are several other factors to consider, but for now, it is sufficient to think of the interest rate as the expected inflation rate plus an additional charge for the perceived risk and opportunity costs.

The interest rates that banks set are in principal guided by the interest rates paid on US Treasuries (which provide the means for the US government to borrow money from investors all over the world and are viewed as safe investments). The development of interest rates on such Treasuries thus reflects very well the development of interest rates in general. Here's how the interest rate of 10-year Treasury Notes has changed since 1962:

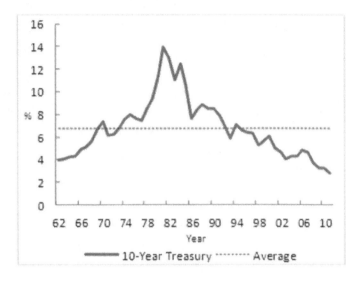

Note that this picture doesn't show the real (inflation-adjusted) interest rate. If the interest rate is high but the inflation rate is as well, your gains might not as great as implied by the interest rate. More on that later in the chapter.

2. Simple Interest

Simple interest is ... well, simple. Or at least simpler than other forms of charging interest. The formula requires three inputs: the principal P (this is the money loaned or borrowed), the interest rate i (expressed as a decimal number) and the time t (expressed in years). The total interest received can be calculated using:

$I = P \cdot r \cdot t$

You loan a friend P = 1000 $ for a period of t = 1.5 years. He agrees to pay i = 5 % = 0.05 interest. This means you'll get:

I = 1000 $ · 0.05 · 1.5 = 75 $

in interest. Or in other words: after the 1.5 years are up, he will pay you back 1075 $ in total.

Your bank agrees to loan you P = 2500 $ from the first July till the first December 2015. Because they perceive the risk of default as relatively high, they will charge i = 9 % interest. What is the amount of interest to be paid?

First we need to express the time span in years. We have 31 days for July, 31 days for August, 30 days for September, 31 days for October and 30 days for November. This makes 153 days, which is 153 / 365 = 0.42 years. So the interest is:

I = 2500 $ · 0.09 · 0.42 = 94.5 $

Which principal P will earn I = 400 $ interest for a credit with an interest rate i = 7 % and a duration of t = 2 years? Inserting all the given data results in:

400 $ = P · 0.07 · 2

400 $ = P · 0.14

Dividing both sides by 0.14 will gives us the desired result:

P = 400 $ / 0.14 = 2857.14 $

----

3. From Simple to Compound Interest

This section is only to present the train of thought that leads from simple to the compound interest. The formulas for compound interest can be found in the next section.

Imagine you loan a bank the principal P = 10000 $ at an interest rate of i = 5 %. This is the amount of interest you would receive, given the duration t of the loan:

t = 1 year

$\rightarrow$ I = 10000 $ · 0.05 · 1 = 500 $

t = 2 years

$\rightarrow$ I = 10000 $ · 0.05 · 2 = 1000 $

t = 3 years

$\rightarrow$ I = 10000 $ · 0.05 · 3 = 1500 $

As you can see, the interest grows linearly with the duration of the loan. For each additional year, you get an additional 500 $, which is just 5 % of the principal 10000 $. In other words: each year the interest rate is applied to the principal. How could that be any different?

Consider this: At the end of the first year, you'll receive an interest payment in the amount of 500 $. This means that your bank statement will now read 10000 $ + 500 $ = 10500 $. So why not apply the interest rate to this updated value? This would lead to an interest payment of 10500 $ · 0.05 = 525 $ for the second year instead of just 500 $.

Continuing this train of thought, at the end of the second year your bank statement would read 10000 $ + 500 $ + 525 $ = 11025 $. Again we would rather have the interest rate applied to this updated value instead of the unchanging principal. This would result in an interest payment of 11025 $ · 0.05 = 551.25 $ for the third year.

For comparison, here's what the final pay out would be for the simple interest plan:

10000 $ + 500 $ + 500 $ + 500 $ = 11500 $

And this is what we would get with the "not simple" interest plan, where we apply the interest rate to the updated amounts instead of the principal:

10000 $ + 500 $ + 525 $ + 551.25 $ = 11576.25 $

The latter is called compound interest. It means that we include already paid interests in the calculation of next year's interest, which leads to the amount received growing exponentially instead of linearly.

More E-Books by Metin Bektas

Great Formulas Explained

In this book you will find some of the greatest and most useful formulas that the fields of physics, mathematics and economics have brought forth. Each formula is explained gently and in great detail, including a discussion of all the quantities involved and examples that will make clear how and where to apply it. On top of that, there are plenty of illustrations that support the explanations and make the reading experience even more vivid. The book covers a wide range of topics: acoustics, explosions, hurricanes, pipe flow, car traffic, gravity, satellites, roller coasters, flight, conservation laws, trigonometry, equations, inflation, loans, and many more. Volume II is available under the title "More Great Formulas Explained".

The Book of Forces

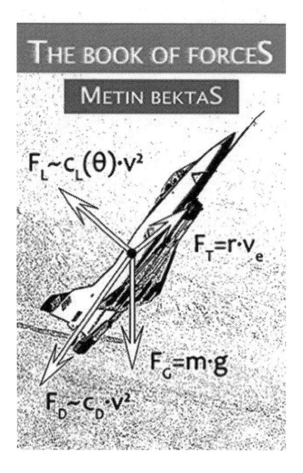

Forces make the world go 'round - literally. This book provides a quick and easy-to-understand introduction to the quantity force and an overview of the many types of forces that shape our universe. Besides enlightening and down-to-earth explanations, you'll find plenty of detailed exercises demonstrating how the concepts and formulas can be applied to real-world situations. Knowledge of high school algebra is sufficient to follow the calculations. For more information, check out the table of contents.

Physics! In Quantities and Examples

This book is a concept-focused and informal introduction to the field of physics that can be enjoyed without any prior knowledge. Step by step and using many examples and illustrations, the most important quantities in physics are gently explained. From length and mass, over energy and power, all the way to voltage and magnetic flux. The mathematics in the book is strictly limited to basic high school algebra to allow anyone to get in and to assure that the focus always remains on the core physical concepts.

Copyright and Disclaimer

Copyright 2015 Metin Bektas. All Rights Reserved.

This book is designed to provide information about the topics covered. It is sold with the understanding that the author is not engaged in rendering legal, accounting or other professional services. The author shall have neither liability nor responsibility to any person or entity with respect to any loss or damage caused or alleged to be caused directly or indirectly by the information covered in this book.

The book is for personal use of the original buyer only. It is exclusive property of the author and protected by copyright and other intellectual property laws. You may not modify, transmit, publish, participate in the transfer or sale of, reproduce, create derivative works from and distribute any of the content of this book, in whole or in part.

The author grants permission to the buyer to use examples and reasonably sized excerpts taken from this book for educational purposes in schools, tutoring lessons and further training courses under the condition, that the material used is not sold or given away and is properly cited.

Made in the USA
Middletown, DE
07 November 2017